# Emerging Technologies in Clean Energy 2019



Topical Conference at the 2019 AIChE Spring Meeting
and 15th Global Congress on Process Safety

New Orleans, Louisiana, USA
31 March - 4 April 2019

ISBN: 978-1-5108-8679-7

# Emerging Technologies in Clean Energy 2019

Topical Conference at the 2019 AIChE Spring Meeting
and 15th Global Congress on Process Safety

New Orleans, Louisiana, USA
31 March - 4 April 2019

AIChE
The Global Home of Chemical Engineers

**Printed from e-media with permission by:**

Curran Associates, Inc.
57 Morehouse Lane
Red Hook, NY 12571

**Some format issues inherent in the e-media version may also appear in this print version.**



Printed by Curran Associates, Inc. (2019)

For permission requests, please contact AIChE
at the address below.

AIChE
120 Wall Street, FL 23
New York, NY 10005-4020

Phone: (800) 242-4363
Fax: (203) 775-5177

www.aiche.org

**Additional copies of this publication are available from:**

Curran Associates, Inc.
57 Morehouse Lane
Red Hook, NY 12571 USA
Phone: 845-758-0400
Fax: 845-758-2633
Email: curran@proceedings.com
Web: www.proceedings.com

# TABLE OF CONTENTS

# 550379 Determination of Effective Transport Parameters from Particle-Resolved CFD Simulations for a Simplified Fixed-Bed Reactor Modeling

Multitubular fixed-bed reactors are widely used in the chemical industry, esp. in the field of heterogeneous catalysis. Each module consists of a fixed-bed with a low tube-to-particle diameter ratio *N* and is characterized by a heterogenous packing morphology that lead to a strong interplay between local flow effects and heat and mass transfer within the system.

Particle-resolved computational fluid dynamics (CFD) is a first-principle modeling approach that has been proven to be a reliable and predictive tool to analyze the flow, temperature and species field within fixed-beds in a spatially resolved manner as recently reviewed by *Jurtz et al.* However, since this approach is numerically demanding it cannot be applied for the simulation of the whole multitubular system in the foreseeable future. Therefore, simplified models, e.g., a pseudo-homogeneous modelling approach, will still be the model of choice to simulate the full reactor.

These simplified models rely on the knowledge of effective transport parameter, e.g., effective thermal conductivity and dispersion coefficient. Most correlations available to estimate these parameters are either not suitable since they were developed for reactors with a high *N* where the impact of local flow phenomena is not that important and plug-flow assumption can be made or are simply not available for the specific particle shape that is used.

To bridge that gap it will be shown how effective transport parameters can be extracted from particle-resolved CFD simulation. The parameters can than be used in simplified models. The combination of particle-resolved CFD and simplified models is a promising approach to make CFD a design-tool for this kind of applications.

Literature
Jurtz, N., Kraume, M. and Wehinger, G. (2018). *Advances in fixed-bed reactor modeling using particle-resolved computational fluid dynamics (CFD).* Reviews in Chemical Engineering. Published: 2018/02/02. Available at: https://doi.org/10.1515/revce-2017-0059 [Epub ahead of print]

# 549688 Simulation of Structural Effects of Monolith Catalyst on Methane Dry Reforming in a Solar Thermochemical Reactor

Methane dry reforming (MDR) is a promising progress, as it not only solves the issue of greenhouse gas ($CH_4$, $CO_2$) emissions, but also provides appropriate energy storage products (CO, $H_2$) for individual applications or further Fischer-Tropsch synthesis. The introduction of solar energy into MDR process is an attractive approach to store solar energy to chemical fuels with eliminating the limitations in time and space. Since MDR is an endothermic process, concentrated solar energy can provide essential heat to maintain the process operation. Since MDR obeys surface chemistry, the catalyst structures affect the thermal and reactive performances. In this paper, computational fluid dynamic (CFD) numerical analysis is applied to study the effects of lattice structure for the monolith catalyst in a three-dimensional (3D) annular fixed bed sample. Solar radiation is introduced as an energy source boundary condition with considering internal radiation via discrete ordinate method (DOM). A detailed micro-kinetic surface mechanism of Ni based catalyst is applied to simulate the MDR process. The structures of catalysts are further built in the catalyst region of former reported solar partition cavity reactor to investigate the heat, mass transfer performance with energy conversion efficiency in a specific reactor.

# 546546 Mitigating Membrane Degradation for Safer and Reliable, Design and Operation of Proton Exchange Membrane Fuel Cells

One of the most promising technologies for moving towards a hydrogen economy is the proton exchange membrane fuel cell (PEMFC) due to its low operating temperatures, low weight and volume, high power density and short startup time. This technology has begun to be used as in the form of fuel cell electric vehicles (FCEVs). However, a major hindrance towards making PEMFCs competitive with hydrocarbon-based combustion engines is the higher costs owing to lower durability of these systems. Specifically, the reliability of membrane exchange assembly (MEA) has been identified to be the key issue contributing towards overall lower durability of these systems. This decrease in reliability is attributed to membrane degradation leading to the overall PEFMC system degradation. Apart from contributing to decreased performance and higher cost, PEMFC degradation has been identified through fault tree analysis as one of the root causes that can lead to hydrogen leak and subsequent explosion.

The consequences resulting from hydrogen leak and explosion in a PEMFC system can be disastrous, especially from the perspective of vehicular application. Therefore, to mitigate the risk associated with explosion in PEMFC system and to achieve the required levels of performance, it is essential to study the degradation of PEMFC membrane in depth. This study deals with understanding the fundamentals of PEMFC degradation through modeling using computational fluid dynamics (CFD). More importantly, the degradation mechanism is quantitatively related to the probability of explosion of the PEMFC system using fault tree analysis. Therefore, an important inference from this study will be the design and operation alternatives for PEMFC that mitigate membrane degradation leading to improvement in both performance and safety of the system.

An interesting feature of this study will be the experimental validation of the derived conclusions from the modeling study. A small scale PEMFC system will be set up and the design and operation alternatives derived from modeling study will be validated. The experimental validation will mainly involve varying the type of membranes in the PEMFC cell, flow rates of hydrogen and oxygen gas, voltage applied and mode of operation (air or pure oxygen mode). The degradation of membranes will be achieved by externally treating them with Fenton's reagent (at varying concentrations) and the degraded membranes will be tested in the PEMFC system as well. The effects of these variations will be measured through the current generated by the PEMFC cell.

# 551269 Fabrication and Characteristics of Ni Doped Titania Nanotubes By Electrochemical Anodization

A semiconductor as $TiO_2$ can be used in photocatalytic processes among several applications. One of them is to produce electricity from solar energy, this is very interesting because it can be used to satisfice the electrical energy requirements on the operation of electrochemical cells. Its band gap makes it very interesting in the photoelectrodes manufacturing used in photoelectrochemical cells for hydrogen production, a clean and environmentally friendly fuel. During this process pairs hole – electron are photogenerated, the efficiency can be improved if charge recombination is avoided, also if the energy absorption into the visible spectrum is increased and its specific surface area is increased.

The synthesis of 1D titanium dioxide nanostructures, such as nanotubes, makes possible to produce more efficient photoelectrodes for solar energy to hydrogen conversion. In essence, this is because it increases the charge transport rate, decreasing recombination options. However, its principal constraint is to be mainly sensitive to UV range, which represents less than 8% of solar radiation that reaches earth's surface. One of the alternatives to modifying the $TiO_2$'s band gap and improving its photoactivity under visible light irradiation is to dope the nanotubes with transition metals. This research wants to use nanotechnology in order to produce titanium dioxide nanotubes arrays (TNTs). Their morphology will be controlled during anodization of titanium foils and nickel be deposited onto the TNTS with the purpose to increase the absorption spectrum of the visible region.

This option requires fabricating efficient nanostructured photoelectrodes with controlled morphology and specific properties able to offer a suitable surface area for metallic doping. Hence, currently one of the central challenges in photoelectrochemical cells is the construction of nanomaterials with a proper band position for driving the reaction while absorbing energy over the VIS spectrum. This research focuses on the synthesis and characterization of Ni doped $TiO_2$ nanotubes for improving its photocatalytic activity in solar energy conversion applications. Initially, titanium dioxide nanotubes (TNTs) with controlled morphology were synthesized by two-step potentiostatic anodization of titanium foil. The anodization was carried out at room temperature in an electrolyte composed of ammonium fluoride, deionized water and ethylene glycol. Consequent thermal annealing of as-prepared TNTs was conducted in the air between 450 °C - 550 °C. Afterwards, the nanotubes were superficially modified by nickel deposition. Morphology and crystalline pase of the samples were analyzed by SEM, EDS and XRD before and after nickel deposition. Determining the photoelectrochemical performance of photoelectrodes is based on typical electrochemical characterization techniques. Also, the morphological characterization associated electrochemical behavior analysis were discussed to establish the effect of nickel nanoparticles modification on the $TiO_2$ nanotubes. The methodology proposed in this research allows using other transition metal for nanotube surface modification.

# 551328 Ferrite Based Two-Step Thermochemical Water Splitting Cycle: Solar Reactor Efficiency Analysis

Production of chemical fuel such as $H_2$ by using solar energy provides a promising path for solar energy conversion and storage. The energy density of $H_2$ is equal to the summation of the energy densities of oil, gas and coal. $H_2$ can be used directly as a transportation fuel and can be utilized to produce electricity via fuel cell operation. In addition, $H_2$ can be mixed with the CO produced from captured $CO_2$ to make syngas, which can be further converted into liquid transportation fuels.

Ferrite based thermochemical splitting of $H_2O$ is considered as one of the promising options for the production of $H_2$. A variety of ferrites such as Ni-ferrite, Co-ferrite, Zn-ferrite, Mn-ferrite, Ni-Zn-ferrite, Ni-Mn-ferrite, and others are investigated heavily towards solar thermochemical $H_2$ generation. Several researchers investigated the effects of synthesis methods, water splitting temperature, thermal reduction temperatures, different reactor arrangements, addition of grain growth inhibitors (mixing and coating approach) and others on ferrite based water-splitting cycle. In addition to the experimental investigations, to compare with other metal oxide based thermochemical cycles, it is highly essential to estimate the solar to fuel conversion efficiency of the ferrite based thermochemical $H_2O$ splitting cycle. In this paper, a thorough thermodynamic scrutiny of solar $H_2$ production via ferrite based $H_2O$ splitting cycle is performed using HSC Chemistry 7.1. Effect of partial pressure of $O_2$ in the inert gas ($N_2$), thermal reduction temperature and water splitting temperature on equilibrium compositions and thermodynamic efficiency of the ferrite based $H_2O$ splitting cycle is investigated.

# 551329 Metal Oxide Based Solar Driven Two-Step Methane Reforming and Water Splitting Cycle

$H_2$ production via metal oxide based solar thermochemical splitting of $H_2O$ is considered as one of the alternatives for the gasification and reforming of fossil fuels, pyrolysis and reforming of biomass, ethanol decomposition, and others. Various metal oxide based thermochemical cycles such as zinc oxide cycle, tin oxide cycle, iron oxide cycle, ferrite cycle, ceria cycle, perovskite cycle, and others have been investigated towards the solar production of $H_2$ via water splitting reaction. The major limitation associated with this process is the requirement of a very high temperature for the reduction of the metal oxide. This issue can be addressed if the metal oxide based $H_2O$ splitting process is coupled with the $CH_4$ reforming process.

Recently, metal oxide (MO) based two-step $CH_4$ reforming and $H_2O/CO_2$ splitting cycle receiving much attention for co-production of syngas and $H_2$. As explained above, in first step, the MO is reduced using $CH_4$ as the reducing agent producing synags (combined $H_2$ and CO). The reduced MO can be re-oxidized using $H_2O$, $CO_2$, or combined stream of $H_2O/CO_2$ for the production of $H_2$, CO, or syngas. In this study, the equilibrium thermodynamic analysis is performed to study the effect of $CH_4$/metal oxide ratio on equilibrium compositions and the temperatures required to achieve higher levels of fuel production. Likewise, the solar reactor efficiency analysis is conducted to determine the solar-to-fuel energy conversion efficiency of this cycle. The investigated metal oxide is further synthesized and characterized using various analytical techniques. The synthesized metal oxide is tested towards $CH_4$ reforming and $H_2O/CO_2$ splitting reaction using a thermogravimetric analyzer (coupled with GC-MS). Effects of $CH_4$ concentration, reforming and splitting temperatures on the fuel production is also explored. The obtained results are compared with previous thermochemical cycles and will be presented in detail.

# 545460 Connecting Chemical Process Design Economics and Entropy Generation

Systematic analysis of chemical process systems with the two steady-state balance equations from the First and Second Laws of Thermodynamics provides insights about both energy efficiency and property models when entropy generation is the indicator of irreversibilities[1-4]. Consistent evaluations of overall processes and of subsections yield reliable comparisons for energy requirements of process alternatives, indicate the impacts of potential changes, and check property model and calculational consistencies.

TEPET is a standardized methodology specifically developed for techno-economic evaluation of alternative fuels manufacture[5]. This approach enables qualitative and quantitative statements regarding the technical and economic feasibility of fuel synthesis concepts, including identification of appropriate fuel production concepts within predefined framework conditions. Specifically, it avoids uncertainties in results of previous studies for assessing and comparing feasibilities of alternatives for fuel manufacturing that would arise from significant differences in applied methodology, level of detail, and key assumptions for economic factors and market prices. The inputs to the software are output files from AspenPlus® simulations, which are heat integrated to minimize utility costs by designing an optimized heat exchanger network based on results from pinch-point analysis.

This presentation describes results from a case study to relate entropy generation to process design economics. AspenPlus® has available process simulation files for the manufacture of fuel grade methanol from syngas feed. Entropy generation and economic analyses have been done for different processes with the same feed and product bases to reveal their differences in irreversibilities and costs. A relationship between entropy generations and cost has been established for these systems. Changes in cost from process improvements suggested by reducing entropy generation have also been obtained.

References:


1. J.P. O'Connell, "Chemical process systems analysis using thermodynamic balance equations with entropy generation", Comp. & Chem. Eng. 2017, **107**, 3-15.
2. J.P. O'Connell, "Chemical process systems analysis using thermodynamic balance equations with entropy generation. Revaluation and extension", Comp. & Chem. Eng. 2018, **111**, 37-42.
3. J. P. O'Connell, "Detailed Examination of Energy Flows and Entropy Generation in Low Pressure Binary Distillation Columns", Ind. Eng. Chem. Res., 2018, **57** (32), 11186–11194
4. J. P. O'Connell, "Analysis of Chemical Process Systems with Entropy Generation", AIChE Annual Meeting, Pittsburgh, PA, paper 421d.
5. F. G. Albrecht, D. H. König, N. Baucks, R-U. Dietrich, "A standardized methodology for the techno-economic evaluation of alternative fuels – A case study", Fuel, 2017, 194, 511–526.

# 549937 Optimizing Catalyst Free-Biodiesel Production with Super-Critical Ethanol and $CO_2$ Co-Solvent Using Response Surface Methodology


**ABSTRACT**

Fossil-based methanol production for biodiesel generation is not considered renewable compared to biodiesel produced from ethanol generated from agricultural biomass via fermentation. In this study, biodiesel production from corn oil using super-critical ethanol and $CO_2$ as a co-solvent is investigated under reaction conditions of 250-350 °C/17MPa with a residence time of 10-30 minute and ethanol to oil ratio of 15-35. Furthermore, the co-solvent pressure was varied between 10-50 bars. In this study, response surface methodology was used to identify the optimum values for biodiesel transesterification yield. Results identified the controlling factors as the reaction temperature, reaction time, $CO_2$ pressure plus second order effects including the temperature with reaction time for biodiesel yield. The highest biodiesel yield of 94% was achieved at 275 °C, 20:1 ethanol to oil ratio, and co-solvent pressure of 40 bar for 25-minute reaction time. The order of significance for reaction parameters for biodiesel yield was reaction time > $CO_2$ pressure > reaction temperature. The process $R^2$ and adjusted $R^2$ were 0.96 and 0.92 respectively.

# 550121 Catalytic Hydrothermal Liquefaction of Food Waste Using Inexpensive Catalysts

A recent DOE study showed that more than 13.6 million dry tonnes of food waste is produced every year in the United States[1], most of which is disposed of in landfills. Globally, there is an estimated 1.3 billion tonnes of food disposed of in landfills annually[2]. Therefore, food waste represents a significant organic resource which is inexpensive, has a high energy density and has the potential to be converted into drop-in transportation or heating fuels[3]. In addition, repurposing food residues helps divert a significant percentage of material from landfills and hence reduces the greenhouse gas emissions produced by the subsequent biodegradation of the organic waste.

Hydrothermal liquefaction (HTL) is a promising method of the conversion of organic waste and biomass streams with high water content, into a carbon rich bio-oil, along with bio-char and residual aqueous phases[4]. However, organic compounds partitioning into the HTL aqueous phase reduce the oil recovery and require costly water purification steps. Therefore, in order to reduce the water-soluble HTL organics, carbon-carbon and other coupling reactions are desirable to increase the product molecular weight, hence the oil phase solubility. Catalysts can promote such coupling reactions. In the literature, both homogenous and heterogeneous catalysts are utilized with HTL to increase the bio-oil yields and energy recovery[5-8]. Recently, we reported the use of $CeZrO_x$ catalyst for HTL of food waste[8]. $CeZrO_x$ catalyzed HTL showed significant improvement to the bio-oil yield and energy recovery compared to non-catalyzed and $Na_2CO_3$-catalyzed HTL, while reducing the organic content of the aqueous phase. Using model compounds, it was shown that $CeZrO_x$ promotes carbon-carbon coupling aldol reactions. In addition, $CeZrO_x$ exhibits hydrothermal stability at 300°C. However, this is a relatively expensive catalyst material containing the rare earth metal cerium.

In this study, we evaluated the catalytic viability of inexpensive acid and base solid oxide catalysts, including red mud (Bayer process waste), red art clay and fly ash (from coal combustion power plants) using batch HTL reactions on a representative consumer food waste at 300 °C and 20.7 MPa. HTL bio-oil products were evaluated for their energy content using higher heating values and total organic content of the aqueous phase product was determined using total organic carbon (TOC) analyzer. The results show that red mud and red clay improved the total oil recovery relative to non-catalyzed HTL reactions by reacting water-soluble organics.

**Acknowledgements**

This work was funded by a Department of Energy SBIR (Grant Number DE-SC0015784). Carla Roma, Caroline Murphy, Lawrence Valeros and Joseph Esposito supported the experimental work on this project.

**References**

[1] U.S. Department of Energy. Biofuels and Bioproducts from Wet and Gaseous Waste Streams: Challenges and Opportunities; U.S. Department of Energy: Washington, DC, USA, 2017.

[2] Food and Agriculture Organization of the United Nations, http://www.fao.org/save-food/resources/keyfindings/en/

[3] Sanjib Kumar Karmee, Renewable and Sustainable Energy Reviews, 53 (2016) 945–953.

[4] S.S. Toor, L. Rosendahl and A. Rudolf, Energy 36 (2011) 2328-2342.

[5] L. Nazari, Z. Yuan, S. Souzanchi, M. B. Ray and C. Xu, Fuel 162 (2015) 74–83.

[6] Zhu, Y.; Biddy, M.J.; Jones, S.B.; Elliott, D.C.; Schmidt, A.J. Appl. Energy 2014, 129, 384–394.

[7] Ross, A.B.; Biller, P.; Kubacki, M.L.; Li, H.; Lea-Langton, A.; Jones, J.M. Fuel 2010, 89, 2234–2243.

[8] Alex R. Maag, Alex D. Paulsen, Ted J. Amundsen, Paul E. Yelvington, Geoffrey A. Tompsett and Michael T. Timko, Energies 2018, 11, 564, 1-14.

# 549573 Carbon-Footprint Analysis for Conversion of CO2 to Alkyl Carbonates

Life Cycle Analysis (LCA) is an essential first step for developing the energy efficiency, alternate energy and carbon-management technologies. Globally CO2 capture and utilization (CCU) is actively pursued as part of the carbon-management program. E3Tec has developed a process for conversion of CO2 to dimethyl carbonate (DMC) and performed detailed C-footprint, which is a first step to full implementation of LCA. DMC and its derivatives are high-value specialty chemicals with major applications in manufacturing polycarbonates, as low VOC solvent, in lithium-ion batteries, and as an intermediate in manufacturing polyurethane. DMC is also evaluated as diesel fuel additive to reduce pollution. An Excel-based C-footprint model was linked to ASPEN Plus® process analysis via a Process Block Diagram (PBD). The PBD which clearly defined the Inside Battery Limit (ISBL) and Outside Battery Limit (OSBL) components and subsystems. C-footprint of the ISBL process was based on the ASPEN Plus® simulation, while that of OSBL components was based on the literature database. This C-footprint analysis facilitated the $CO_2$ conversion process by identifying components and/or subsystem that could be further developed to enhance the overall energy efficiency and hence reduce C-footprint. The analysis is being extended for an integrated process of $CO_2$ capture and conversion to alkyl carbonates.

# 551137 The Development of a Pressure Swing Adsorption Model to Evaluate Metal Organic Frameworks

Metal organic frameworks (MOFs), a diverse class of crystalline materials that are defined by metal ions coordinated to organic ligands, are a promising candidate for $CO_2$ capture. The advances in this area of reticular chemistry over the last 25 years has lead metal organic frameworks to become a playground for chemists to design unprecedented porous materials that can be used for reversible $CO_2$ capture. However, identifying MOFs for reversible $CO_2$ adsorption presents a major challenge because of the number of competing phenomena within a MOF that contribute to effective capture and regeneration within an adsorption column.

In this work we present a pressure swing adsorption model that quantifies the potential of a specific MOF to act as an adsorbent within a packed bed adsorber. The model allows for the user to input experimentally determined values for the adsorption isotherm and effective diffusivity constants of the MOF. This allows for the MOF's complex microstructure to be translated into a set of macroscopic adsorption and diffusion parameters that determine saturation and desaturation times of a pressure swing adsorber. The model assumes plug flow through an adsorption column filled with M spherical particles of a specific MOF. The packed bed of length L is then divided into i stages of size L/N with M/N identical particles. Diffusion into a porous sphere with appropriate boundary conditions is assumed to be the driving force of $CO_2$ uptake into the MOF particles.

We implemented this model to predict the behavior of an adsorption column packed with the MOF known as zeolitic imidizolate framework-8 (ZIF-8). Preliminary results for an adsorption column on the length scale needed for installation into a vehicle suggests that approximately 2.5% of all $CO_2$ input into the system can be adsorbed over the saturation time of the filter. A saturation time on the order of two hours is achieved for a flow rate of 0.144 g $CO_2$ per hour. These results show that our model offers a route towards evaluating the performance of specific MOF geometries in large scale adsorption processes. Our model presents a unique solution to determine which MOFs are worth investing resources into for further development as a large scale $CO_2$ capture technology.

# 549010 Systematic Synthesis Framework for the Development of $CO_2$ and $H_2$ sinks: A P-Graph Approach

Carbon dioxide ($CO_2$) emissions are increasing at a significant rate, in order to meet the energy demands of a growing population. To curb the detrimental effects of these emissions, whilst providing for the energy requirements of the population, an energy source that utilizes $CO_2$ should ideally be introduced into our infrastructure. This research utilizes a P-Graph superstructure approach, to encompass a wide array of $CO_2$ reaction pathways that produce methanol and dimethyl ether – both of which have proven to be prominent in the energy industry. Reaction pathway possibilities were simulated using Aspen Plus software. 85 Possible case studies were then analyzed from an environmental and economical perspective, in order to determine the most feasible pathway to produce the aforementioned products. The use of a $CuO/ZnO/Al_2O_3$ catalyst at 240°C and 80 bar pressure, was found to be the most feasible option to produce methanol. The most feasible scenario for the production of dimethyl ether, on the other hand, is to produce methanol over the aforementioned catalyst, and then further dehydrate the methanol over an Al/Si catalyst at 400°C and 25 bar pressure. This presentation aims to detail the systematic framework used in the P-Graph analysis, when arriving at these two reaction pathways.

# 551324 Synthesis of Ceria Based Redox Materials Using Co-Precipitation Method for Solar Thermochemical Application

Solar driven thermochemical splitting of $H_2O$ and $CO_2$ using metal oxide based redox reactions is considered as a promising new technology for the production of syngas, bypassing the utilization of fossil fuels. Syngas i.e. a mixture of $H_2$ and CO is the feedstock for the production of liquid transportation fuels such as gasoline, diesel, and jet fuel via the Fischer Tropsch process. The production of solar syngas via thermochemical splitting of $H_2O$ and $CO_2$ is an attractive green process. Among the various metal oxides investigated in past, ceria based redox materials are considered as promising once due to their excellent redox reactivity and high fuel production capacity. In this investigation, the ability of various ceria based redox materials towards splitting of $CO_2$ was estimated experimentally. The ceria based redox materials were prepared via co-precipitation method and tested towards the $CO_2$ splitting reaction using a high temperature TGA, and packed bed reactor set-up. The pre-reaction and post-reaction characterization of the ceria based redox materials was performed using different analytical methods such as powder XRD, BET surface area analyzer, scanning and transmission electron microscopy, and others. Obtained experimental results will be presented and compared with previously published literature.

# AUTHOR INDEX

**AIChE**
120 Wall Street, FL 23
New York, NY 10005-4020

ISBN 978-1-5108-8679-7

9 781510 886797

CURRAN ASSOCIATES INC.
proceedings.com